Class A Foam for
Structural Firefighting

Investigated by: Jeff Stern
J. Gordon Routley

This is Report 083 of the Major Fires Investigation Project conducted by Varley-Campbell and Associates, Inc./TriData Corporation under contract EME-94-C-4423 to the United States Fire Administration, Federal Emergency Management Agency.

Homeland
Security

Department of Homeland Security
United States Fire Administration
National Fire Data Center

U.S. Fire Administration Fire Investigations Program

The U.S. Fire Administration develops reports on selected major fires throughout the country. The fires usually involve multiple deaths or a large loss of property. But the primary criterion for deciding to do a report is whether it will result in significant "lessons learned." In some cases these lessons bring to light new knowledge about fire--the effect of building construction or contents, human behavior in fire, etc. In other cases, the lessons are not new but are serious enough to highlight once again, with yet another fire tragedy report. In some cases, special reports are developed to discuss events, drills, or new technologies which are of interest to the fire service.

The reports are sent to fire magazines and are distributed at National and Regional fire meetings. The International Association of Fire Chiefs assists the USFA in disseminating the findings throughout the fire service. On a continuing basis the reports are available on request from the USFA; announcements of their availability are published widely in fire journals and newsletters.

This body of work provides detailed information on the nature of the fire problem for policymakers who must decide on allocations of resources between fire and other pressing problems, and within the fire service to improve codes and code enforcement, training, public fire education, building technology, and other related areas.

The Fire Administration, which has no regulatory authority, sends an experienced fire investigator into a community after a major incident only after having conferred with the local fire authorities to insure that the assistance and presence of the USFA would be supportive and would in no way interfere with any review of the incident they are themselves conducting. The intent is not to arrive during the event or even immediately after, but rather after the dust settles, so that a complete and objective review of all the important aspects of the incident can be made. Local authorities review the USFA's report while it is in draft. The USFA investigator or team is available to local authorities should they wish to request technical assistance for their own investigation.

For additional copies of this report write to the U.S. Fire Administration, 16825 South Seton Avenue, Emmitsburg, Maryland 21727. The report is available on the Administration's Web site at http://www.usfa.dhs.gov/

U.S. Fire Administration
Mission Statement

As an entity of the Department of Homeland Security, the mission of the USFA is to reduce life and economic losses due to fire and related emergencies, through leadership, advocacy, coordination, and support. We serve the Nation independently, in coordination with other Federal agencies, and in partnership with fire protection and emergency service communities. With a commitment to excellence, we provide public education, training, technology, and data initiatives.

 Homeland Security

ACKNOWLEDGMENTS

The United States Fire Administration appreciates the help of the following persons that provided information for this report:

Sven Carlson	FoamPro Corporation
Sam Duncan	U.S. Army Tank-Automotive Command, Ft. Belvoir (VA)
Deputy Chief Ken Jones	Fairfax County Fire & Rescue Department (VA)
Chief Bill May	Westlake Fire Department (TX)
Doug Miller	Task Force Tips/KK Products
Eric Peterson	National Fire Protection Research Foundation (MA)
Captain Gary Pope	Fairfax County Fire & Rescue Department (VA)
Kevin Roche	Phoenix Fire Department (AZ)
Assistant Chief Rick White	Nashville Fire Department (TN)
Underwriters Laboratories	

TABLE OF CONTENTS

Class A Foam For Structural Firefighting
December 1996

Reported by: Jeff Stern
              J. Gordon Routley

OVERVIEW

The increasing use of class A foam systems by urban and suburban fire departments for structural fire suppression has created a demand for information on this technology. While class A foams have been used extensively by wildland and rural fire departments, their application to structural fire suppression is a recent trend. This report discusses the use of class A foaming agents in conjunction with water for fire suppression (*conventional or nozzle-aspirated class A foam systems*); it also provides additional information on the use of class A foam agents with water and compressed air (*compressed air foam systems, or CAFS*)

Many fire departments have conducted their own field testing of class A foam to evaluate its effectiveness in structure fires. Several departments have attempted to adapt class A foam equipment that was originally developed for wildland firefighting for structural firefighting operations. Many benefits of class A foam have been reported, including quicker fire extinguishment, faster overhaul time, less damage to buildings, and reduced fatigue on firefighting personnel due to quicker mop-up after the fire is out. Additional advantages reported with CAFS include the ability to maneuver attack lines easily and the ability to extend available water supply for a longer period of time. Exposure protection is greatly enhanced with class A foam.

This report begins with a general overview of nozzle-aspirated class A and compressed air foam systems. It then discusses hands-on evaluations by several fire departments that are currently using class A foam systems in structural fire suppression or wildland/urban interface fire protection. These departments were contacted to provide a candid overview of their experience with class A foams and/or CAFS. In all cases, these departments view class A foam and CAFS as additional tools which increase the efficiency and effectiveness of their fire suppression operations. Reported advantages and disadvantages in the use of class A foams and CAFS in structural firefighting, both from field experience and from recent fire protection literature, are included in this report.

Recent studies conducted by Underwriters Laboratories in cooperation with the National Fire Protection Research Foundation and the U.S. Army Fort Belvoir Research, Development, and Engineering Center compared the use of plain water, nozzle-aspirated class A foam, and CAFS for the extinguishment of class A fires. The experiments showed that nozzle-aspirated class A foams and CAFS generally extinguished the experimental fires more quickly and used less water than plain water extinguishing methods; though in some cases plain water was shown to be equally effective.

The use of compressed air foam systems (CAFS) for structural firefighting was previously evaluated by the United States Fire Administration in Technical Report 074 of the Major Fires Investigation Project in 1993. That report, "Compressed Air Foam For Structural Fire Fighting: A Field Test; Boston, Massachusetts," highlighted the experimental use of a class A compressed air foam system by the Boston Fire Department's Engine Company 37 in 1992-93. The field test indicated some operational advantages provided by CAFS and encountered some shortcomings in retrofitting the existing apparatus. The Boston test showed that more information is needed to be gathered to determine the extinguishing capabilities and conservation of water supply when using CAFS in an urban environment.

Summary of Key Issues

Issue	Comments
Strategy and Tactics	Departments reported no need to change basic fireground strategy or tactics when using nozzle-aspirated class A or CAFS compared to plain water.
Equipment reliability	Few failures of equipment were reported. Proportioner microprocessors may need to be better isolated from water.
Water supply conservation	Water conservation appears to be a significant advantage of CAFS. The reduced flow rate effectively doubles the capability of tank water. Water conservation is less significant with nozzle-aspirated class A foam.
Water damage reduced	Experimental studies indicate that less water is used to extinguish fires when foam is used. The departments contacted for this report did not document reductions in water damage in urban/structural firefighting.
Extinguishment	Departments report quicker fire control and extinguishment with both nozzle-aspirated class A foams and CAFS, compared to plain water.
Overhaul	Departments report reduced damage during overhaul and less time spent on overhaul and mop-up.
Hoseline management	Firefighters report that CAFS lines are much easier to handle than water lines due to decreased weight and nozzle reaction. Nozzle operators on CAFS lines must be aware of increased initial nozzle reaction from the build-up of compressed air at the nozzle before the line is opened. No difference was noted with nozzle-aspirated class A foam.
Costs of equipment	Departments reported equipment costs up to $5,000 per unit for nozzle-aspirated class A foam systems. CAFS units may cost up to $40,000 per vehicle.
Costs for training	Most departments did not quantify the cost of training their personnel to operate class A foam systems.
Costs for foam	Class A foam costs approximately 1/10 of the cost of class B foams per gallon of foam produced. The cost of the foam concentrate was not a limiting factor for any of the departments contacted.
Safety	Safety may be enhanced through reduced fatigue during fireground operations, due to quicker extinguishment and reduced overhaul time. Eye protection and rubber gloves are necessary when handling foam concentrate due to its harsh detergent properties.

I. INTRODUCTION TO CLASS A FOAM SYSTEMS

HISTORY OF FOAM USE ON CLASS A FIRES

The use of foam to fight class A fires was first evaluated in the 1930's. Early studies showed that foams applied to class A fuels could suppress fires more efficiently than plain water in most cases; however, the available foam concentrates were expensive and could only be mixed at high concentrations. They were not a cost effective means of extinguishing fires. The experimental use of CAFS to produce foam dates back to the 1940's.

The State of Texas began using foaming additives for brush and wildland firefighting in the 1970's. Foam was produced by early compressed air foam systems that used an air compressor mounted on board the fire apparatus to aerate a solution of water and foam concentrate. These early CAFS units were mechanically troublesome but very effective at combating grass and brush fires. Several hundred of these units were eventually developed in Texas and remain in widespread use today.

In the 1980's, synthetic foam concentrates were developed for use on class A fires that could be applied at low concentrations of 0.1 to 1.0 percent. These new concentrates made the use of foam a cost effective means of combating fires because smaller amounts of foam concentrate could be used to make effective foam. The use of class A foams for wildland firefighting expanded in the 1980's and 1990's. Recent generations of CAFS and nozzle-aspirated class A foam concentrate induction systems have been more reliable than earlier models. Nozzle-aspirated class A foams and CAFS units have been developed that can deploy foam from firefighters' backpacks, from brush and fire engines, and from airplanes and helicopters.

Gradually, as fires in wildland/urban interface areas have increased in severity and cost, the urban fire service has been exposed to the use of class A foams. Several urban departments have experimented with different types of foam systems for structural firefighting and some have deployed front-line apparatus with class A foam systems as an added tool in their firefighting arsenals.

TYPES OF FOAM DELIVERY SYSTEMS

Class A foams consist of a mixture of water, foam concentrate, and air. The composition of the foam depends on the proportion of the three components. The two most common methods for producing class A foam are nozzle-aspirated foam systems and compressed air foam systems (CAFS).

Conventional or Nozzle-Aspirated Class A Foam Delivery Systems (NAFS)--In nozzle aspirated class A foam delivery systems, the water and foam concentrate are mixed together via an educator or by a mechanical proportioning device to create foam solution. In most class A systems, this occurs on the discharge side of the water pump. The foam solution is delivered to the nozzle where it is aerated to form the class A foam.

Foam concentrate can be pre-mixed into the apparatus water tank; however, this has several drawbacks, including possible corrosion of the tank, stripping of the fire pump lubricants, and contamination of the foam solution with rust and scale from the tank, which may adversely affect the ability to produce the type of foam desired.

Nozzle-aspirated class A foams are low energy foam systems which only use the hydraulic energy supplied by the water pump to propel the foam stream.

Compressed Air Foam Systems (CAFS)--The compressed air foam system consists of a water pump, an air compressor, and foam concentrate proportioning device.

In CAFS, the foam solution (water and concentrate) is mixed on the discharge side of the pump, similar to nozzle-aspirated class A foam delivery systems. Compressed air is then introduced into the mixture. This aerates the foam prior to distribution through the hoselines.

Since the product flowing through the hoselines is finished foam, it consists of large amounts of air and a reduced amount of water. Therefore, the hoselines weigh much less and are more flexible than plain water hoselines. They are more easily advanced and maneuvered than water or nozzle-aspirated class A foam hoselines.

The CAFS hose stream is projected a longer distance than a plain water stream under the same pressure, due to the reduced mass and added energy of the compressed air. In wildland operations CAFS are often operated without a nozzle on the end of the delivery hose. A shut-off nozzle with a straight bore tip is normally used for structural firefighting. The compressed air can build up behind a closed nozzle, which can cause a severe nozzle reaction if the air is not bled off properly by the nozzle operator before fully opening the nozzle.

CAFS is considered a high energy foam delivery system because the hydraulic energy of the pressurized water is combined with the pneumatic energy of pressurized air, both to aerate and to propel the foam.

Some disadvantages exist with the use of CAFS. The addition of the air compressor complicates the pump operator's job by adding several additional steps to produce compressed air foam. Slug flow may occur if not enough concentrate is mixed into the foam solution, leaving the fire suppression crew with an ineffective stream. If the compressor fails, the hoseline will still deliver plain water or foam solution, but a nozzle must be placed on the line to create an effective stream (if the line is being operated without a standard nozzle).

CAFS units can deliver plain water (with the foam eductor off and air compressor shut down) or nozzle-aspirated class A foam (without using the air compressor) in addition to the compressed air foam. Class B foams can also be produced with CAF systems. Experiments have shown that class B CAFS streams are highly effective in providing exposure protection from class B fires.

FOAM CHARACTERISTICS

Class A foams has physical characteristics that vary depending upon the method of production. The characteristics depend upon the concentration of the foam solution, type of concentrate used, hose length, nozzle type, and means of aeration. In most cases, the pump operator can control the type of foam produced. Foam is generally characterized as wet, fluid, or dry.

Wet foam--Wet foams are characterized by smaller bubbles, less expansion because less air has been introduced, and fast drain times. Wet foams are generally good for initial fire suppression, overhaul, and penetration into deep-seated fires.

Fluid foam--Fluid foam has been characterized as having the consistency of watery shaving cream. Fluid foam tends to have medium to smaller bubbles and moderate drain times. Fluid foam is good for direct attack, exposure protection, and mop-up operations.

Dry foam--Dry foam has a high expansion ratio and the consistency of shaving or whipped cream. The dry foam is very fluffy and consists mainly of air. Dry foams have slow drain times and hold shape for a long period of time. Dry foam is particularly good for exposure protection because of its ability to cling to vertical surfaces for extended periods.

HOW CLASS A FOAM WORKS

Class A foam is 99 percent water. The foam increases the efficiency of water as a fire extinguishing agent.

Heat--The primary method of fire extinguishment with class A foam is the absorption of heat energy from the fire to convert water molecules to steam. This is the same process that makes plain water an effective extinguishing agent. The heat-absorbing properties of water are not changed by the addition of foam; however, the foam bubbles provide a greater surface to mass area for the water, which may allow the liquid water to convert to steam more rapidly.

Fuel--Class A foams act as an insulating blanket on class A fuels, protecting exposed and recently extinguished surfaces from the heat of a fire to inhibit ignition (or re-ignition). The surfactant action of class A foam reduces the surface tension of water, allowing the water to penetrate and soak into fuels, instead of running off. This property is particularly significant in reducing overhaul requirements and penetrating tightly packed materials.

Oxygen--The foam blanket may help to form a vapor barrier between the burning fuels and oxygen.

FOAM PROPORTIONING DEVICES

Proportioning devices inject the foam concentrate into water to make foam solution. Several types of proportioning devices can be used with class A foam concentrates. The types of proportioners listed here inject foam at the discharge side of the pump, which is the most common method of installation.

Eductors--Foam eductors use the venturi effect of flowing water to draw foam concentrate out of a container and into a hose stream. These devices are identical to class B foam eductors, but regulate the concentrate at a much lower percentage. Eductors are limited to one hoseline per eductor, and the percentage of concentrate is usually limited to a specific setting within a limited range. Eductors are the least expensive induction system next to batch mixing the foam solution in the water tank. If the eductor is not pre-piped, firefighters must connect the eductor into the hoseline and into the concentrate container when they arrive on the fire scene.

Eductors are simple devices with few or no moving parts, and do not require a power supply. They are limited by specific water flow and water pressure requirements, usually must be within 150 feet of the nozzle to operate effectively, and introduce a significant amount of friction loss into the hoseline. Eductors are widely used for class B foams because of their low cost and maintenance, and because class B fires are relatively rare.

Balanced Pressure Bladders--Bladder proportioners contain foam concentrate in a flexible bladder which is contained within a water tank connected to the water pump. Water is by-passed from the discharge line into the tank, which squeezes the bladder and causes the foam concentrate to be discharged into the water stream. The rate of release is controlled by the rate that water is allowed to enter the tank and displace the contents of the bladder.

Bladder type proportioners do not require external power sources. The foam delivery must be interrupted in order to refill the bladders with foam concentrate.

Concentrate Pumps--Mechanical or electric pumps can also proportion foam concentrate into the water stream. Several manufacturers produce packaged proportioning systems that mix water and foam concentrate at a rate set by the pump operator. This is accomplished through mechanical means (such as a venturi) or by electronic controls (flow meter), depending upon the type of unit. Concentrate pumps require an additional power supply and are more complex than other proportioning methods, but they allow the pump operator to select the percentage of concentrate over a wide range (usually between .1 and 1 percent for class A foams) to control the characteristics of the foam produced. These pumps are often more accurate over a wide range of flow rates than other proportioning devices.

Some of these systems can work with two or more foam concentrate tanks, allowing the pump operator to choose between a class A foam concentrate and a class B foam concentrate, depending upon the type of fire encountered. This could be useful for departments in an urban setting that already use class B foams and desire to incorporate class A foams as an additional extinguishing agent.

Concentrate pumps allow for the tanks to be refilled without having to interrupt the flow of foam, which is an advantage over eductor or bladder proportioning systems.

NOZZLES

Nozzle-aspirated class A foam can be used with standard firefighting nozzles. Special foam nozzles are not required.

Combination nozzles will help aerate the foam solution, forming a very wet foam with little expansion. The combination nozzles can also be used with CAFS; the combination nozzle will act to strip away the large bubbles formed in the foam, leaving a wet foam stream.

Smooth bore nozzles can be used with nozzle-aspirated class A foam or CAFS. With CAFS, a large orifice smooth bore nozzle will help straighten and project the foam stream. The larger nozzle size is used because the hose is delivering finished foam to the tip, which expands rapidly when it leaves the nozzle. Some wildland firefighting units apply CAFS with no nozzle on the hoselines.

Specialized foam nozzles are available for class A foam. These are often used in the wildland or wildland/urban interface setting, or for exposure protection.

Most of the departments contacted for this report use traditional smooth bore or combination nozzles to deliver class A foams, either via nozzle-aspirated foam systems or CAFS. The Nashville (TN) Fire Department uses combination nozzles to achieve 3:1 expansion rate on their nozzle-aspirated class A foam systems. The Fairfax County (VA) Fire and Rescue Department uses 3/4-inch smooth bore nozzles when flowing CAFS.

TESTING AND APPROVAL OF FOAM CONCENTRATES

Testing--Most of the class A foam concentrates are used primarily by wildland firefighting organizations. The U.S. Forest Service and the Department of Agriculture require class A foam concentrates to be approved for use on wildland fires. The concentrates must pass a series of tests for product stability and storage, corrosion, health and safety, and operational evaluations. NFPA 298 (1989), *Standard on Foam Chemicals for Wildland Fire Control* also addresses class A foam concentrates, but does not set any performance measures for structural firefighting.

U.S. Government Approved Class A Foam Concentrates--The following class A foam concentrates have received interim approval for use on ground fire engines from the U.S. Forest Service at the time of this publication. An updated list of approved foams can be received by contacting the National Interagency Fire Center (NIFC) in Boise, Idaho. Some of these foam concentrates have also been approved for use in aerial applications (helicopter or air tanker). Departments can contact the NIFC for more information.

Class A Foam Concentrate	Mix Ratio
Ansul Silv-Ex	0.1 - 1.0 %
Fire-Trol FireFoam 103	0.1 - 1.0 %
Phos-Chek WD 881	0.1 - 1.0 %
Fire-Trol FireFoam 104	0.1 - 1.0 %
Angus ForExpan S	0.1 - 1.0 %
Procap B-136	0.1 - 1.0 %
TCI Fire Quench	0.1 - 1.0 %

Storage of Foam Concentrates--Class A foam concentrates should be stored according to the manufacturer's guidelines. Concentrates should generally be stored in their original containers, either 55-gallon drums or 5-gallon cans. Apparatus concentrate tanks should be constructed of polyethylene, polypropylene, fiberglass, or other plastic composite material. The foam concentrate will cause degradation of steel, aluminum, and some stainless steel tanks, which could lead to damage or malfunction of the foam proportioning equipment.

II. SPECIAL CONSIDERATIONS FOR USE OF CLASS A FOAMS

ENVIRONMENTAL CONSIDERATIONS

In general, class A foams are much more environmentally friendly than most class B film-forming foams, which will not biodegrade well and often must be cleaned up as toxic waste after use. Care should be taken to prevent spills of concentrate into waterways and watershed areas, because aquatic life is sensitive to foaming agents. The use of foam in wildland firefighting has proved that foam has little effect on forests soils and plant life due to its ability to rapidly degrade.

Federally approved class A foams are tested for their ability to biodegrade into inert components within an established period of time. For approval by the U.S. Forest Service, 50 percent of foam must biodegrade within 28 days. Most foams biodegrade within 14 to 30 days. Foams are also tested for their toxicity to certain types of fish and marine life.

The National Wildfire Coordinating Group advises that the following guidelines be followed to prevent damage to aquatic environments when using foams. The group advises that these precautions be taken near domestic reservoirs and domestic water supplies as well.

- Train all personnel in the potential problems of introducing foam concentrates into bodies of water.

- Locate foam mixing and loading areas where there is minimal contact with natural bodies of water.

- Avoid spills at mixing and loading areas, especially when located near live streams.

- Exercise caution when using foams in watersheds where fish hatcheries are located.

- Leave a 100- to 200- foot buffer zone between the area where foam is used and the high water line.

Departments that use class A foams in an urban or suburban setting should be aware of these protective measures to prevent adverse environmental impacts. The impact of foam in public sewage systems has not been documented.

PERSONAL SAFETY CONSIDERATIONS

Foam concentrates are harsh detergents which can irritate the skin, causing dryness, cracked skin, and bleeding. Diluted foam solution should have little or no effect on personnel.

Proper precautions should be taken when handling foam concentrates to prevent injury. Personnel that handle concentrates should wear goggles and rubber gloves to prevent skin and eye irritation. Long-sleeved shirts and long pants are recommended. Rubber boots are also recommended, as the concentrate can soak through leather boots quickly. It is recommended that spills of concentrate be soaked up with absorbent rather than flushed with water (which will create a lot of foam). Personnel should have access to emergency eye wash equipment should concentrate splash into the eyes.

When using foam, personnel should be aware of slippery surfaces. This should require little adjustment for personnel in a structural environment, where water run-off tends to leave surfaces slippery, anyway. Few departments reported any problems with slippery conditions.

Personnel must also avoid drinking water from tanks where foam concentrates have been introduced. Warning signs should be posted to prevent thirsty firefighters from drinking water out of these tanks.

Government approved class A foams are tested for human toxicity. Several tests are conducted, including determining the acute oral limit, the acute dermal limit, primary dermal irritation, primary eye irritation, and the acute inhalation limits. Results of these tests should be available from the manufacturer and listed on the material safety data sheets shipped with the foam concentrate.

OTHER SPECIAL TACTICAL USES

While useful in general day-to-day operations, class A foam has the potential to have a great advantage in certain fire situations when compared to plain water.

Conflagrations--Departments facing large, uncontrolled fires such as wildland/interface fires, or fires that have occurred after earthquakes, could deploy nozzle-aspirated class A foam or CAFS units as a means of protecting exposures, confining and extinguishing fires. The ability of CAFS units to engage in rapid extinguishment and overhaul could greatly increase the capabilities of companies combating large fires that may result from wildfires, earthquakes, or civil disturbances.

The special ability of CAFS to conserve water supplies is an advantage, especially for independent task force units facing conflagration situations where hydrant systems have been destroyed.

Unstable structures--Fires that occur in unstable or unsafe buildings could be fought from a greater distance by using the long reach of CAFS foam streams. Crews could remain at a safe distance outside of the collapse zone. Theoretically, additional weight from water would be reduced with the use of foam, lessening the chance for a collapse.

Lightweight construction--The rapid and enhanced fire suppression capability of nozzle-aspirated foam systems and CAFS could improve fire suppression when fighting fires in modern, lightweight construction or trussed-roof structures.

III. FIRE DEPARTMENT EVALUATIONS AND EXPERIENCE

Several fire departments that currently use class A foam for structural fire attack or wildland/urban interface fire suppression provided their experiences with evaluating, implementing, and using class A foam.

NASHVILLE FIRE DEPARTMENT (TENNESSEE)

The Nashville Fire Department serves a major city and surrounding jurisdiction covering over 533 square miles. The department initiated the use of class A foams for structural firefighting operations in 1991. The department utilizes nozzle-aspirated class A foam systems on engine companies.

A committee was formed to evaluate class A foam technology in March of 1991. The committee decided in favor of nozzle-aspirated class A foam over a CAFS system due to the simplicity of use and lower costs. The department determined that eight nozzle-aspirated class A systems could be placed in service for the same price as one CAFS system. In January of 1992, five engines were outfitted with class A foam systems. They have since been placed on 17 engines. Nashville's class A foam systems were retrofitted to Emergency One, FMC, Mack, and International engines by the fire department shop. The costs per engine were between $3,500 and $3,700.

Nashville involved members from all areas of the department in their initial evaluation, planning, and implementation of class A foam. The training academy developed a four-hour class limited to the various uses of class A foam, operations of foam equipment, and troubleshooting the mechanical systems. The department provided training to three companies at a time over a 20-day period, until all personnel were trained. The course was designed to keep explanations as simple as possible; the chemistry of class A foam as a firefighting agent was not discussed.

There was some initial resistance to the implementation of class A foam. Union representatives helped from the outset to allay fears that this would become a means to reduce staffing on units. Some initial resistance among experienced chief officers and older departmental members was also encountered, but this was overcome as the advantages of foam use were experienced during fire-ground operations.

The department has averaged about 300 gallons of foam concentrate use per engine per year. They generally purchase foam in a yearly 5,000-gallon batch for between $9 and $10 per gallon. The department has used several different manufactured brands of class A foam without difficulty. All foam used has been approved by the U.S. Forest Service and meets NFPA Standard 298 where applicable.

The department uses 1-3/4-inch hose with a combination tip nozzle which aspirates the foam solution, creating a foam expansion ratio of about 3:1.

The Nashville Fire Department has had a very positive experience with the use of class A foam. Firefighters in Nashville use foam in virtually every situation where plain water was previously applied. Firefighters describe quicker fire knockdown times and improved visibility due to rapid dissipation of the products of combustion.

Some problems have also occurred when some firefighters failed to discern between class A foam concentrate and class B AFFF foam concentrate canisters. Severe damage to foam system components occurred in instances when firefighters, by mistake, added class B foam concentrate to a class A foam concentrate tank. (The mixing of the different concentrates caused the concentrated AEFF to congeal, gel, and clog the foam tank and system, requiring the entire system to be removed and cleaned. This problem has been reported by several departments.) A simple solution suggested by the Nashville Fire Department was for the manufacturers to color code class A and class B foams by colored container. Currently, all foam concentrate containers look similar, whether they contain class A or class B foam concentrate.

Several microprocessors on the foam proportioners have also failed, requiring replacement, a relatively simple procedure.

Some other disadvantages have also been reported, including increased reports of hotter steam conditions as the fire is knocked down (by 25 or 30 degrees). At the time of this report, Nashville was attempting to evaluate the need for rapid ventilation of structures when attacking with class A foam, due to these reports. It has not been determined if this increased temperature is real or a perception; it may relate to firefighters going more deeply and aggressively into hot areas of the building than they did prior to the use of foam. Other departments have not noted this particular problem. (It should be noted that experiments in the use of class A foam conducted by the Underwriters Laboratories for the National Fire Protection Research Foundation and for the U.S. Army did not reveal any increase in temperatures during fire suppression with foam over plain water.) The department reports that no firefighter injuries 'have been attributed to the use of class A foam.

Nashville's nozzle-aspirated foam hoselines weigh the same as plain water lines (Nashville does not use CAFS). The nozzle aspirated system does not have the easier maneuverability of a CAFS line, but if the foam system should fail, the firefighters still have a fully charged plain water hoseline.

Most Nashville Fire Department personnel have felt that class A foam provides benefits for structural firefighting. The class A foam engines have performed well during structural fireground operations. Firefighters and incident commanders have reported quicker fire extinguishment. The department believes that they have been able to confine and suppress larger fires using class A foam. Commanders report that attack and overhaul operation time has been reduced, reducing fatigue on their companies and allowing them to return to service more quickly. This has provided Nashville with an increase in company availability for EMS calls, which constitute 83 percent of the department's emergency responses. The ability of companies to quickly return to service is seen as a significant advantage, and an improvement in the level of service they can provide.

Nashville was successful in implementing the use of class A foam by including all ranks of the department in evaluating the foam system. It is especially important to include line firefighters and union representatives in the evaluation of this technology.

Nashville has used class A foam on several hundred structure fires in the last few years, and they report quicker or equal knockdown time with foam, in almost all cases, as well as quicker overhaul times and units returning to service more quickly. The department has not documented any savings in water damage or water use.

The Nashville Fire Department plans to continue the use of class A foam.

PHOENIX FIRE DEPARTMENT (ARIZONA)

The Phoenix Fire Department has used nozzle-aspirated class A foam on a limited basis for four years, and compressed air foam systems for two years. The department uses class A foam for wildland firefighting, interface firefighting, and structural fire suppression.

Phoenix has two engines equipped with nozzle-aspirated class A foam systems, and two engines with CAFS. Four brush units also have class A foam systems. The department also has a telesquirt unit with a combination system capable of flowing either class A or class B foam. All of the existing units have been installed by fire department mechanics. Six new pumpers with factory-installed CAFS are on order.

Members of the fire department who have worked with class A foam have stated that the foam provides a definite advantage, especially in deep-seated fires in piles of wood or tires.

The first foam systems used in Phoenix were found to be unnecessarily complex for structural firefighting; newer, simpler systems have been received more positively by the crews. Phoenix developed general awareness training for firefighters and command level officers on the use of class A foams. Training for pump operators has been provided on an individual basis in the stations where the units are deployed.

No tactical changes have been identified when using class A foam in structural firefighting. Foam is used at the discretion of the company officer or incident commander. Phoenix currently has no formal written policy for class A foam use at structure fires.

Like Nashville, Phoenix firefighters have provided anecdotal information about faster fire knockdown and quicker overhaul when using foam. Phoenix has not attempted to determine if less water damage is occurring when foam is used.

The department plans to continue expanding its fleet of class A foam units in the future.

FAIRFAX COUNTY FIRE AND RESCUE DEPARTMENT (VIRGINIA)

The Fairfax County Fire and Rescue Department (FCFRD) had one CAFS unit in service at the time of this report. The unit was originally built for experimental use and evaluation in conjunction with the Fort Belvoir (U.S. Army) Fire Department. This CAFS equipment was retrofitted to an existing pumper at the fire department shop. This was a lengthy and difficult process for the fire department mechanics, but enabled the department to conduct field evaluations of the CAFS system at reduced costs.

After the evaluation phase, the CAFS unit was placed in service to enhance structural fire attack in an area characterized by low water supply, long second-in company response times, and some of the large and expensive homes.

The Fairfax Fire and Rescue Department reported having a positive experience with the CAFS unit. The foam system was thought to reduce firefighter fatigue through diminished suppression and overhaul times. Fairfax hopes this will contribute to reduced firefighter injuries during the overhaul stages of fires in the future. Key advantages are seen in the ability of CAFS to maximize the effectiveness of a limited water supply, effectively doubling the capabilities of the water tank on the CAFS engine. Due to the long response times into the areas where the first CAFS unit has been deployed, this is seen as a major advantage. Other advantages include the lighter and easier mobility of the CAFS attack lines, enabling firefighters to easily handle up to 2-1/2-inch hoselines.

Fairfax County had major problems retrofitting the CAFS on an older engine, which was not designed for the unit's components. Several problems were encountered with the original retrofitted unit (Boston had a similar experience). The evaluation report produced by the U.S. Army after the Fort Belvoir tests recommends against retrofitting existing fire apparatus with a CAFS, due to the complexity and size of the components that must be added.

The department recently purchased a new CAFS engine to replace the retrofitted unit. Fairfax County estimates that the factory-installed system on their newly purchased fire engine was about $35,000 over the cost of a non-CAFS engine; however, they estimate that the CAFS unit will prove to be 60-100 percent more effective than a plain water engine, effectively giving them a fire suppression capability equivalent to two fire engines. The estimated increase in efficiency is based on their experience during the Fort Belvoir evaluation period. The retrofitted CAFS engine will be reassigned to a station in another limited water supply response area. The department is currently discussing the purchase of a third CAFS engine by a volunteer fire company within the county.

Station personnel were trained in the uses of the CAFS. Little resistance has been encountered. One chief officer considers the use of this technology a possible paradigm shift in fire suppression. The department has not yet had a large fire at which the value of the CAFS unit could be put to the test for evaluation by the whole department.

The Fairfax County Fire and Rescue Department is currently formulating operational guidelines for the use of the CAFS. Like the other departments, Fairfax has noted no new tactical considerations. The crews were pleased with the weight reductions in the hoselines, though they noted that the increased energy stored in the nozzle results in a high nozzle reaction when the nozzle is first opened.

Evaluators in the Fairfax County Fire and Rescue Department point out that CAFS is an effective tool in enhancing their fire suppression capability. The ability to coat and protect exposures with foam, to extend the reach of hose streams with the added energy of the CAFS, and to extend the available water supply are seen as key advantages in the decision to employ CAFS in fireground operations.

WESTLAKE FIRE DEPARTMENT (TEXAS)

Class A foam is employed by the Westlake Fire Department in Travis County, Texas, by both nozzle-aspirated and compressed air foam systems. The department is a combination career and volunteer department with three fire stations protecting 19 square miles of residential and commercial property in the suburbs of Austin. The department faces a severe wildland/urban interface fire risk and employs class A foam as a tool in both interface fire suppression and structural firefighting.

The Westlake Fire Department had a CAFS unit for several years, but it was in poor condition and rarely used. In 1994, the department began to add updated foam equipment and to increase training of all personnel. The department currently has several pieces of apparatus equipped to use class A foam. A brush unit was converted to CAFS with a 100-cfm air compressor and a 200-gpm pump. The department also converted a military truck into a large CAFS unit with a 200-cfm compressor and a 500-gpm pump. Department planning currently calls for the retrofit of two existing fire engines with nozzle-aspirated class A foam inducting systems. Specifications are also being prepared for a new CAFS engine.

Several advantages in class A foams have been cited by the Westlake firefighters, including the wetting agent action of the foam, better sheeting and absorption of the class A foam into fuels. Also, the visibility of the foam lets them see the areas where it has been applied in contrast to plain water.

Westlake also noted the lightweight and easy hose management characteristics and faster knockdown that other departments have experienced.

The mess associated with foam production and the need to shut down operations to refill concentrate bladders on some types of proportioning systems were seen as disadvantages.

The department uses equipment from several different manufacturers. Their apparatus includes Westex, Ford, Sutphen, KME, and others, with Robwen proportioners and FoamPro proportioning systems. The Westlake Fire Department estimates that it has invested over $25,000 in its small CAFS unit and almost $40,000 in its large CAFS unit.

Firefighter training has been conducted locally and through courses offered by the State of Texas. Firefighters have been enthusiastic about the increased use of the foam systems in the department. The firefighters feel the additional capabilities provided by foam are advantageous in their daily operations. Department guidelines call for the use of class A foam or CAFS at the discretion of the company officer.

Westlake noted the increase in nozzle reaction of CAFS as well as problems with slug flow when the CAFS attack lines are charged prior to the introduction of the foam concentrate. The introduction of the foam concentrate (which allows the air and water to mix) at the proper time will prevent the slug flow effect. Slug flow can also occur when too little concentrate is introduced into the foam solution.

Westlake officials indicated that poor quality foam proportioners may have increased their costs over the long term due to inaccurate foam proportioning, which has lead to excessive use of concentrate. The quality of the foam induction equipment should be thoroughly evaluated by departments considering purchase of foam systems. Air compressors, a key component of the system, should also be researched. The Westlake Fire Department saved additional costs by building their own air-aspirating foam nozzles out of PVC pipe, rather than purchasing manufactured nozzles.

Westlake suggests that reductions in water damage in a structural setting may be dependent upon the fire stream management by the suppression crews. Heavy application of foam may actually contribute to some water damage by soaking into areas that would not normally be affected, due to its increased penetrating ability.

Westlake plans on continuing the use of class A foam.

IV. SUMMARY OF ADVANTAGES AND DISADVANTAGES

ADVANTAGES

The fire departments that are using class A foam and recent literature list many advantages in the use of class A foam over plain water, including the following.

1. **Class A foam allows faster fire suppression and extinguishment than with plain water.**

   Firefighters report that fire knockdown often occurs more quickly with class A foam. This has been substantiated in both actual fireground operations and experimentation. Firefighters and incident commanders have reported some cases where foam outperformed plain water lines while, in other cases, the foam performance was considered equal to water. There were no reports of inferior performance.

   Tests conducted by the Underwriters Laboratories for the National Fire Protection Research Foundation compared plain water streams against class A foam in a series of comparative tests and came to the same conclusion. Underwriters Laboratories conducted additional tests for the U.S. Army with similar results.

2. **Class A foam increases efficiency and conservation of water supply.**

   The increased efficiency per gallon of water is most evident with CAFS. The difference in effectiveness per gallon of water is estimated in the literature as high as 5 to 10 times over plain water for some applications. A CAFS engine with a 500-gallon water tank would have the equivalent fire suppression capability of a vehicle with a 2,500- to 5,000-gallon water tank.

   Water supply is conserved because less gpm is needed per hoseline. The experimental use of CAFS in the City of Boston in 1992 showed that a 1,000-gpm pumper with a 700-gallon water tank, could operate a single CAFS 1-3/4-inch attack line from the tank water for approximately 10 minutes before needing to secure an additional water supply. With plain water the 700-gallon tank would only be able to supply a 1-3/4-inch attack line for three to four minutes. The Boston firefighters estimated that the CAFS attack line had about the same capability to knock down the fire as the 1-3/4-inch line using plain water. This could provide a tactical advantage in situations where establishment of water supply is delayed.

   Reduced water use was noted in several tests. With foam, less water will remain for run-off and associated water damage from firefighting operations and overhaul in a structure fire. The use of class A foam tends to reduce the amount of water that is needed to control and overhaul all types of fires, particularly where densely packed or compressed fuels are involved. However, the impact appears to be most significant in rural and wildland/urban interface firefighting, where the water supply may be limited to the capacity of the tank on the attack vehicle.

   In theory, the reduced use of water could also be advantageous in lessening the contribution of fire suppression activities to building collapse, because the applied foam would weigh less than a comparable amount of water. This has not yet been documented in actual fireground operations, nor in experimental studies.

3. **Class A foam can be produced at a relatively low cost.**

    Class A foam concentrates are proportioned at rates between 0.1 percent and 1.0 percent. (This compares with class B foam concentrates which are proportioned at 3 percent to 6 percent.) At a rate of 0.3 percent, 1,000 gallons of class A foam can be produced with only 3 gallons of class A foam concentrate and 997 gallons of water; the estimated cost of the concentrate would be about $30 (assuming a cost of $10/gallon of concentrate). An equivalent amount of class B foam at 3 percent would require 30 gallons of class B foam concentrate and 970 gallons of water: at a cost of about $300 dollars (assuming the same cost per gallon of foam concentrate).

    One fire department contacted for this report estimated that the cost of their class A foam concentrate was probably offset by the savings in their use of diesel fuel resulting from reduced operating time on the fireground.

    The lower cost of class A foam can reduce the cost of training for class B fires. Some departments that could not previously afford to use class B foams for training are currently using class A foam to simulate foam application. In addition, class A foams are biodegradable and more environmentally friendly than class B foams, so less clean up is required after training.

4. **Class A foam forms a protective blanket.**

    Like water or class B foam, class A foam extinguishes a fire by cooling, but it also has a secondary effect of separating the fuel from its oxygen supply by forming a vapor barrier. This blanket also insulates unburned fuels and exposures from radiant heat or direct flame impingement. This property is particularly effective in protecting exposures and preventing re-ignition after a fire has been knocked down.

5. **Foam is visible during and after application.**

    Class A foam, especially CAFS, is visible during and after application. The visible foam allows firefighters to determine when an area has been adequately covered and when additional coverage is necessary. This is especially useful in wildland/interface firefighting situations where structures must be protected along a large fire front, or in urban situations where an exposure building is threatened by radiant heat or direct flame impingement.

6. **Foam clings to most surfaces and protects exposures much longer than plain water.**

    The ability of class A foam to cling to most surfaces provides advantages in reducing water run-off, helping to reduce water damage and aiding fire extinguishment. The clinging foam solution also aids in the protection of exposures, particularly vertical surfaces and sloped areas. This effect is greatest with CAFS, but can be significant with nozzle-aspirated class A foams as well. Foam can be applied to an exposure and left for a period of time before a reapplication is necessary. (Plain water generally requires a constant flow of water to provide exposure protection.) In addition, the reduced surface tension of foam-enhanced water allows it to penetrate more deeply into class A fuels.

7. **CAFS attack lines are lighter than plain water hoselines.**

    Attack lines that are used to deliver compressed air foam are significantly lighter and easier to handle than plain water handlines, because the product inside the hose is mostly air. The line weighs approximately half the weight of a regular hoseline of the same diameter.

The reduced weight and increased maneuverability can reduce firefighter fatigue and stress. Firefighters can easily handle larger diameter CAFS lines. (Nozzle-aspirated class A lines weigh approximately the same as plain water lines, because they contain the same amount of water.)

8. Foam use may help to preserve evidence of fire cause.

The wetting agent property of class A foams allows them to penetrate and extinguish deep-seated fires in combustible class A materials. This reduces the amount of manual overhaul necessary in the fire area. The fire scene may be better preserved for investigators to determine the fire cause because there is less disruption for overhaul and less damage caused by the impact of the hose streams.

The class A foam eventually evaporates or can be removed to allow for inspection and investigation.

9. Class A foam can be used on flammable liquid fires.

Early tests demonstrated that class A foams may be effectively used on some class B flammable liquid fires, although their relative efficiency as compared to class B foam concentrates has not been documented.

10. Class A foam aids wildland/urban interface attack

Class A and CAFS were originally developed for wildland firefighting and controlling interface fires. Class A foam has been deployed from portable pumps, brush and fire engines, and dropped from aerial tankers and helicopters. The advantage of foam over plain water in the wildland/urban interface settings has been documented over many years.

11. Class A foam may provide long-term cost savings and reduced property damage.

The use of class A foams may lead to long-term cost savings in terms of property saved and resources deployed, over what would have been incurred with the use of plain water alone; however, this has not been conclusively documented.

The quick extinguishment and exposure protection afforded by class A foam and CAFS should lead to decreased total property damage from fires and from fire suppression activities. Departments using foam have documented saving property with foam that they believe could not have been saved using older, plain water firefighting tactics.

12. Firefighter stress and fatigue may be reduced.

The use of class A foam or CAFS may reduce physical stress on firefighters by contributing to faster fire suppression, reduced time to conduct overhaul activities, and faster turn-around time for companies involved in fire suppression activities. This factor is particularly applicable to CAFS, due to the lighter weight and easier maneuverability of the line.

DISADVANTAGES

Class A foam is not without certain disadvantages. Departments evaluating the use of nozzle-aspirated class A or CAF systems must carefully weigh the benefits desired against the costs that may be incurred. Here are some disadvantages reported in the use of nozzle-aspirated class A foam and CAFS.

1. **Initial cost of equipment and training may be substantial.**

   Nozzle-aspirated class A foam and CAFS systems require a considerable initial outlay of costs for equipment, foam, and training. A simple class A foam system consisting of an eductor and foam could cost a few hundred dollars; most class A foam proportioners are in the $2,500-5,500 range. A large compressed air foam system on an engine could cost over $35,000. Departments must also take into account the cost of the concentrate (about $9-10 or more per gallon), as well as the cost to train their personnel. Maintenance costs of the systems should also be considered.

   It is not feasible to document loss control and financial savings in terms of fire suppression efficiency and property saved.

2. **Class A foam concentrate is a corrosive detergent.**

   Like other foam concentrates, the class A foam concentrate is a corrosive detergent which could corrode metal tanks and pump parts. For this reason, most class A foam systems inject the foam concentrate on the discharge side of the pump. Also, the concentrate may be damaging to the paint and finish on fire apparatus.

   Additionally, the concentrate may cause drying and chapping of exposed skin on personnel who handle the concentrates. Personnel handling concentrates should follow safety precautions as outlined on the foam manufacturer's material safety data sheet, including wearing rubber gloves and eye protection. Rubber boots are also recommended, as the concentrate may soak through leather boots.

3. **Long-term environmental impacts are still uncertain.**

   The environmental effects of foam use, especially class A foams, have not been completely determined. Class A foams approved by the U.S. Forest Service are 50-percent biodegradable within 28 days of application. The effects of the foam on humans and wildlife over a long period of time have not been determined. Toxicity data is available from approval tests.

   However, class A foams are considered more environmentally friendly than class B foam, which often must be collected as hazardous material waste after use.

4. **Foam concentrate may cause slip hazards.**

   Firefighters must be aware that foam concentrates may cause a slip hazard if they are spilled, depending upon the surface. Some departments felt the foam created somewhat of a slip hazard beyond plain water, and others did not note any additional hazard.

5. **The effect of foam on fire investigation laboratory tests has not been thoroughly researched.**

   More work is needed to develop techniques for investigation when class A foam has been used, and to educate investigators and firefighters in these techniques. The foam may show up in tests for flammable liquids and accelerants. Additional laboratory tests may be necessary when conducting incendiary investigations to separate chemicals introduced by the class A foam from any chemicals or accelerants that may have been involved in the fire. This is an operational consideration that requires more evaluation and research as foam use becomes more widespread in urban areas.

6. **Firefighters may confuse class A foam with conventional class B foam uses.**

Misidentification of class A or class B foam concentrates has led to damage to expensive foam producing equipment. Class A foam concentrate containers are very similar in appearance to class B concentrate containers. Should firefighters mix class A and class B concentrates, they may cause the product to gel together, severely damaging foam proportioning equipment. This could be rectified by thoroughly training personnel, clearly marking class A and class B foam equipment, and by changing the identification of class A and class B concentrate containers to make them more distinct.

Other problems have been reported with incident commanders who do not understand the differences in class A and B foams. In one reported instance, a CAFS unit was deployed to protect a helicopter landing area during a wildland/urban interface fire because the incident commander did not understand its most efficient use. All fireground commanders should receive training in the tactical uses of class A foams and CAFS.

7. **More possibility for equipment failures.**

Some of the nozzle-aspirated class A and CAFS systems contain complex and computerized equipment. Any additional mechanical equipment creates more points at which a system failure could occur. In CAFS systems, this could potentially compromise the fire stream until the pump operator has the ability to correct the water pressure and the crew puts the nozzle on the line (if necessary); in nozzle-aspirated class A foam systems, the hose would simply deliver plain water should the foam system fail.

Additional procedures and additional equipment logically create more opportunity for mishaps and errors of both a human and a mechanical nature.

8. **Restrictions in CAFS line discharges may reduce gpm flow when using plain water.**

Compressed air foam system (CAFS) lines may have restrictions or baffles built into the pump piping to agitate the foam solution and air to form better quality foam. These restrictions can reduce total gpm flow when the unit is flowing plain water instead of CAFS. One department designated separate, plain water-only discharges to overcome this problem. Some systems have also been designed without these baffles.

V. RECOMMENDATIONS AND CONCLUSIONS

Class A foams have been used for almost two decades in wildland firefighting. The use of class A foams via nozzle-aspirated or compressed air foam systems in structural firefighting is a relatively new technique for building fire suppression. Departments report no need to change basic fire-ground strategy or tactics when using class A foam, and decreased on-scene time for fire units due to rapid extinguishment and reduced overhaul time.

Improvements in the design and manufacture of class A foam delivery systems have increased their mechanical reliability. Departments using class A foam systems today report fewer problems than were encountered several years ago. Most fire department personnel respond positively to the use of these foams during structural firefighting operations. Departments that have included personnel from all levels of the organization in evaluations have had the most success in implementing the use of foam. The final conclusions of this report indicate the following:

1. **Class A Foam is becoming a proven tool for structural fire suppression.**

    Fire departments are increasingly willing to look at alternative technologies that will enhance their operating effectiveness. Class A foam is gaining acceptance as an additional resource for structural fire suppression.

2. **Initial costs may be significant and current data does not establish whether reductions in property losses and improvements in fire suppression efficiency and effectiveness will balance the expenditure.**

    User reports are very positive; however, the cost/benefit analysis is difficult to evaluate without extensive testing.

3. **More research needs to be done to verify reductions in water damage with foam use.**

    No department was able to quantify that reduced water use had actually led to reductions in the cost of water damage where foam was used. Efforts should be made to collect information to validate or disprove this perception.

4. **Departments should include all internal customers in evaluating new technologies like foam.**

    Departments that incorporated members of all ranks and divisions had the best success in assimilating class A foam into the traditional firefighting environment. It is especially important to address the concerns of line firefighters as well as command officers.

5. **Standard operating procedures need to be developed for using class A foams in structural firefighting.**

    Most departments use foam at the discretion of the company officer. Written SOP's or SOG's should be developed as the use of foam becomes more widespread. These procedures should also require Material Safety Data Sheets (MSDS) for the foam concentrate to be posted, and include procedures for storage and safe handling of foam concentrates, as well as recommendations for foam use on the fireground.

6. **More training and education needs to be done with fire investigators to determine where foam helps or hinders fire and arson investigation.**

    Firefighters, fire officers, and investigators should be taught about the affects of class A foam on fire investigations. The influences of foam on laboratory tests need to be determined.

7. **Fireground commanders should be educated in the tactical uses of class A foam.**

    It is important for the fireground commander to have an understanding of the new technologies available so they may be best evaluated and assimilated into department operations.

8. **Manufacturers should consider clearly distinguishing class A and class B foam concentrate containers.**

    Further efforts should be made to prevent firefighters from accidentally mixing class A and class B foam concentrates, which severely damages equipment.

9. **Research should be conducted into additional uses for class A foam, such as in residential sprinkler systems where foam could greatly enhance the suppression abilities given the lower gpm water flow of sprinkler heads.**

    The potential of foam to enhance the fire suppression capability in structural fire suppression systems should be further explored. This could be useful in mobile homes where smaller, self-contained systems could be installed, or in larger commercial, automatic sprinkler systems.

APPENDIX A

UL Report of Class A Foam Tests

(Excerpts reproduced with permission.)

Underwriters Laboratories Inc.®

333 Pfingsten Road
Northbrook, Illinois 60062-2096
(708) 272-8800
FAX No. (708) 272-8129
MCI Mail No. 254-3343
Telex No. 6502543343

REPORT OF

CLASS A FOAM TESTS

PREPARED BY
UNDERWRITERS LABORATORIES INC.
PROJECT 93NK24320/and93NK24320A/NC222

FOR THE

DEPARTMENT OF THE ARMY
BELVOIR RESEARCH, DEVELOPMENT AND
ENGINEERING CENTER

Fort Belvoir, VA

FEBRUARY, 1994

27

EXECUTIVE SUMMARY

Class A foams have been used to fight forest and brush fires for many years. The United States Department of Agriculture (USDA) investigates Class A foams with respect to their toxicity and environmental characteristics. There are no test methods or requirements specified in the National Fire Protection Association (NFPA) Standard for Foam Chemicals For Wildland Fire Control, NFPA 298, to evaluate the firefighting effectiveness of these foams.

Under this research project, wood crib fire and exposure protection tests were conducted to evaluate the firefighting effectiveness of Class A foam hand hoselines as compared to water only. Foam quality tests were also conducted as a part of the research project. These tests were conducted using six Class A foams on the Qualified Products List (QPL) published by the USDA, a UL Listed one percent aqueous film forming foam (AFFF) and water only. Due to the limited number of tests conducted under this investigation, the results were considered inconclusive with respect to quantifying the firefighting effectiveness of Class A foams.

The wood crib fire tests were conducted using Class 20-A wood cribs described in the Standard for the Rating and Fire Testing of Fire Extinguishers ANSI/UL 711. These cribs were designed to be extinguished by a 33 gpm straight stream hoseline applying water only for one minute. For this series of tests, a hand held nozzle set to a straight stream position and fitted with an air aspirating attachment was used at a flow rate of 15 gpm. Class A foam solution concentrations of 0.5 or 1.0 percent were used for all of the tests except those with water only. Except for one of the Class A foam solutions, the results of the wood crib fire tests demonstrated the ability of the Class A foam solutions to extinguish the Class 20-A wood crib. During baseline tests conducted with water only at 15 gpm, the Class 20-A wood crib was not extinguished at the end of the 60 second discharge.

Exposure protection tests were conducted using water only and a Class A foam solution concentration of 0.5 percent. All of the tests were conducted using a hand held air-aspirated nozzle at a flowrate of 1 gpm.

4.0 DISCUSSION AND RECOMMENDATIONS

DISCUSSION:

GENERAL

Due to the limited number of tests conducted under this investigation, the results were considered inconclusive with respect to quantifying the firefighting effectiveness of Class A foams. However, the limited tests did demonstrate the ability of hand hoselines supplied with Class A foam solutions to provide enhanced firefighting performance compared to hand hoselines supplied with water.

WOOD CRIB FIRE TESTS

The results of the wood crib fire tests demonstrated the ability of the Class A foam solutions to reduce the time required to control the fire as compared to water only. During the fire tests conducted with water only, neither wood crib was extinguished as evidenced by visible flaming at the end of the 60 second water application. Fire tests conducted at a 1.0 percent Class A foam solution concentration had the longest reignition times.

EXPOSURE PROTECTION

The results of the exposure protection tests demonstrated the ability of the Class A foam solutions to lengthen the ignition time of a combustible surface as compared to water only at a heat flux value of 50 kW/m². Except for Foam G, the average ignition times of the wood cribs exposed to the Class A foams were as much as 50 percent longer as compared to those exposed to water only at the 50 kW/m² heat flux value.

RECOMMENDATIONS:

Additional research should be undertaken to develop appropriate test procedures and requirements to establish an acceptable level of firefighting performance for Class A foams. Class A foam test requirements contained in NFPA 298 and those developed by the USDA address environmental characteristics only.

There is a need to establish minimum performance criteria for Class A foams to provide a means for evaluating their ability to (1) suppress or control fires, (2) retard the ignition of combustible surfaces exposed to high levels of heat flux and (3) adhere to or be absorbed into both horizontal and vertical surfaces of combustible materials.

The effects of bubble size and generation method would also appear to impact the efficiency of Class A foams which would need to be further researched.

APPENDIX B

Evaluation of NDI Compressed Air Foam System Applied As A Retrofit

(Excerpts reproduced with permission.)

TARDEC

Technical Report

No. 13606

Evaluation of NDI Compressed Air Foam System (CAFS) Applied as a Retrofit

August 1994

By Samuel Duncan

U.S. Army Tank-Automotive Command
Research, Development and Engineering Center
Warren, Michigan 48397-5000

Section III   Feasibility Evaluation

FEASIBILITY OF RETROFITTING CAFS

Retrofitting CAFS to an in-service pumper appears to be too costly in terms of dollars and time; the apparatus is out of service during the retrofit. Three experienced, motivated maintenance personnel with virtually unlimited equipment had a great deal of difficulty over a 6 month period applying the equipment recommended. It might have proved less costly to gut the pumper by removing the pump, tank and the plumbing, including drive train, to rebuild the pumper with the necessary components for CAFS.

There cannot be a recommendation for retrofitting CAFS technology to in-service Army apparatus without a pre-configured kit that can be quickly and efficiently applied to provide CAFS capability.

FEASIBILITY OF CAFS

The CAFS technology was evaluated in a Class B and a Class A scenario. Despite initial equipment difficulties the capability of CAFS generated AFFF to effect exposure protection for fire threatened collapsible fuel tanks was significant. CAFS generated foam in structural fire fighting compared to water proved to be far superior. In all evolutions CAFS proved to be capable of knocking the fire down faster, using less water, reducing the weight of the hose and increasing discharge distance over standard equipment. The foam could be made to stick to overhangs, vertical surfaces such as walls, and to ceilings thereby improving the cooling effect of the water. The CAFS generated foam successfully exhibited all three primary technological characteristics and provided superior fire suppression and protection.

The results of the evaluation are a strong recommendation for CAFS technology, whether for use in TDA fire departments protecting post, camps and stations, municipal fire departments or the Engineer Firefighting Detachments.

Section IV Conclusion

The results of the CRADA support two conclusions. The first is retrofitting CAFS technology to in-service Army fire trucks is not cost effective without a complete, easy-to-install kit. If a kit is developed it should be a "universal" type, capable of fitting the myriad of fire trucks in the Army inventory. No kit exists at this time.

The second conclusion is that CAFS technology provides firefighters with much improved capability to fight fires by increasing the distance of discharge, reducing water requirements and increasing the cooling ability of water by causing the foam to adhere to burning or exposed fuels. Hose line weight is significantly reduced thus mitigating one of the primary physical stressors of fire fighting. Fire trucks could be smaller without losing total firefighting capability. CAFS technology can be built into new trucks for about 15% of the base truck price.

Section V   Recommendations

Based on the results and conclusions of this evaluation, it is the unanimous · recommendation of the project members of the CRADA that CAFS technology would significantly improve the performance of most fire trucks and should be considered in all future fire truck procurements. The technology is simple enough when engineered into the truck at the outset of design, and effective enough in extinguishing fires to be of great value. The performance of CAFS could be improved by additional research to refine or improve the characteristics of CAFS.

NOTES AND OPINIONS

There is no question that additional research is required to bring firefighting into the 21st century. We are still fighting fires in the same manner as when fire was discovered - lots and lots of water. We are not questioning the role of water as the chief agent for fire suppression, rather its effectiveness as it is being used. Conclusive proof from many legitimate sources such as the National Fire Protection Association, Factory Mutual Research Corporation and others, show 80% of the damage in a fire is caused by the massive amounts of water rather than the fire. The deaths and injuries associated with fire incidents are directly attributable to the fire. We must find a better way. We can learn it the hard way on the fire ground, incident by incident; or we can learn through the agent of research.

The CAFS characteristics result in reduced costs and increased safety for any fire department. For the Engineer Firefighting Detachment's worldwide mission, reducing the amount of water required is critical. These firefighting soldiers have the ability to protect and deploy forces but lack appropriate equipment.

The project members are in accord regarding the importance technology must play in fire protection. The fire service, Department of Army or civilian sector, has traditionally been slow to accept change giving rise to the adage "150 years of dedicated service unhampered by progress". Fire departments can no longer rely on the proximity and availability of another engine company when they get into trouble. Fire departments and emergency personnel can no longer rely on unending budget streams, either. They can no longer knock the door down and pummel the contents with hose streams pushing 250 gallons per minute at 125 psi. The handwriting is on the wall - becoming more efficient, effective and safer isn't a better way to do business, it is the only way to stay in business.

Our conclusions, particularly the second, should not be construed to indicate that Army fire departments can operate with less personnel or that fewer firefighters would be required on the fire ground where CAFS equipment is present. Fires in structures designed for living or those that have high occupancy, require four firefighters - one at the pump panel, two on the hose and one to direct the operation and otherwise assist in rescues, hose lays or the myriad of actions that may be necessary to save lives and protect property from fire damage.

There is no question that CAFS reduces water requirements and provides faster knockdown. There is no question that CAFS also prevents reignition as well as initial ignition of exposures. Sadly, there is no question that education of personnel involved in this very special and dangerous field, at all levels, is urgently needed to prevent the loss of one more building, the loss of one more valuable acre of wildland and the loss of one more precious life.

APPENDIX C

NFPRF National Class A Foam Research Project
Technical Report

(Excerpts reproduced with permission.)

NATIONAL CLASS A FOAM RESEARCH PROJECT
TECHNICAL REPORT

STRUCTURAL FIRE FIGHTING - ROOM BURN TESTS
Phase II

Prepared by

William M. Carey, P.E.
Underwriters Laboratories Inc.

NATIONAL
FIRE PROTECTION
RESEARCH FOUNDATION

FIRE RESEARCH

NATIONAL FIRE PROTECTION
RESEARCH FOUNDATION

BATTERYMARCH PARK
QUINCY, MASSACHUSETTS, U.S.A. 02269

Executive Summary

Class A foams have been used to fight forest and brush fires for many years. Recently, municipal fire departments have been using Class A foams to improve the operating efficiency of manual hose streams for structural fire fighting purposes. Phase I of this Research Project involved the conduct of laboratory analysis, wood crib fire, retention and exposure protection tests as described in the National Fire Protection Research Foundation (NFPRF) Report on Class A Foam for Manual Hose Streams dated December, 1993.

This report covers Phase II of the Research Project which involved the conduct of a series of structural fire suppression tests in an 8-by-12-by-8-foot- (2.4-by-3.7-by-2.4-meter-) high enclosure. The enclosure was positioned so that the products of combustion were collected in a calorimeter hood. This permitted measurement of the heat release rate for each test.

Two series of fire tests were conducted in the enclosure. For both series, the walls of the enclosure were fitted with plywood wall paneling having a Flame Spread Index[1] (FSI) of 200 and the ceiling was fitted with tile having an FSI of 25. For the first series of fire tests, a residential sprinkler fuel package as described in the Standard for Residential Sprinklers for Fire Protection Service, UL 1626, and which simulates the upholstered furniture fuel package used in the Los Angeles Residential Sprinkler Tests,[2] was placed in a corner of the enclosure. An opening 5 feet wide by 7 feet (1.5 by 2.1 meters) high was centered in one end of the enclosure for test observation and manual application of the agent.

The fuel package was ignited and the time to reach flashover in the enclosure was recorded. Five seconds after flashover, either plain water, Class A foam or Class A compressed air foam (CAF) was applied until suppression was achieved.

For the second series of tests, the fuel package was changed to an upholstered L-shaped sofa located in the corner of the room. The sofa was ignited in the corner and permitted to burn until flashover occurred. Five seconds after flashover, plain water, Class A foam or Class A CAF was applied until suppression was achieved.

The results of the Phase II fire tests indicated that the use of Class A foam solutions generally reduced the amount of heat released from the fire and damage to the combustibles as compared to plain water. In the Series I fire tests at 5 gpm (18.9 lpm). Class A foam applied using the direct application method took less time and had a lower total heat release from agent application until the rate of heat release was reduced to 500 kW as compared to plain water. No fire tests were conducted with Class A CAF applied using the direct application method. With the agents applied using the indirect method, water and Class A foam had almost identical test results whereas the Class A CAF values were higher. For the Series II fire tests, Class A foam applied at 10 gpm (37.9 lpm) using the

---

[1] Flame Spread Index (FSI) is a fire spread characteristic measured in accordance with the Standard for Test for Surf ace Burning Characteristics, ANSI/UL 723.

[2] Sprinkler Performance in Residential Fire Tests. Technical Report RC8-T-16, Serial No. 22574, Factory Mutual Research Corporation, July 1980.

indirect method took less time and quantity of agent to reduce the rate of heat release to 500 kW as compared to Class A CAF and water only. However, Class A CAF applied at 7 gpm (26.5 lpm) using a direct application method had the shortest time and lowest quantity of agent to reduce the rate of heat release to 500 kW.

It is recommended that additional research be conducted to develop product performance criteria for Class A foams and a method to evaluate specific combinations of Class A foams, proportioning and foam generation methods.

4.0 Discussion and Recommendations

DISCUSSION:

General

The data developed during this series of room fire suppression tests conducted using a single representative Class A foam concentrate generally demonstrated the ability of manual hose streams supplied with Class A foam solutions to provide enhanced structural fire fighting performance as compared to manual hose streams supplied with water only. It should be noted that all tests were performed under laboratory conditions using specific, repeatable test methods and procedures.

Foam Quality

The results of the foam quality tests conducted with the discharge devices indicated that Class A CAF had the highest expansion ratios and longest 25 percent drainage times. It should also be recognized that the standard spray test nozzle used with the Class A foam solution in this Research Project was designed to operate with water only.

Room Fire Suppression Test

During the Series I fire tests employing the UL 1626 residential sprinkler fuel package, the polyether foam was essentially consumed at 100 seconds. There was limited time between flashover, which generally occurred at 80 to 90 seconds, and burnout of the foam, which generally occurred at 100 seconds, for the agent to gain suppression of the fuel package. Based upon a review of the data provided in ILLS. 7 and 8, the following observations were made:

1. Class A foam applied using a direct application method took less time and quantity of agent and had a lower total heat release from agent application until the rate of heat release was reduced to 500 kW than plain water.

2. Using the indirect application method, water and Class A foam had almost identical application times, quantities of agent and total heat release from agent application until the rate of heat release was reduced to 500 kW, whereas Class A CAF values were higher.

For the Series II fire tests, there was sufficient fuel available for the agent to suppress the corner sofa fuel package prior to consumption of the mattresses. Based upon a review of data provided in ILLS. 9-11, the following observations were made:

1. When compared to water only, the test results using Class A foam solutions generally provided for a reduced amount of total heat release from the fire and less damage to the sofa.

2. In general, Class A foam applied at 10 gpm (37.9 lpm) using the indirect method took less time and quantity of agent to reduce the rate of heat release to 500 kW as compared to Class A CAF or water only.

3. Class A CAF applied at 7 gpm (26.5 lpm) using a direct application method had the shortest time and lowest quantity of agent to reduce the rate of heat release to 500 kW.

4. Although direct application of water at >30 gpm (>113.6 lpm) had the fastest suppression time and lowest total heat release and damage, the flow rate was at least three times higher than the flow rate used for the Class A foam room fire tests.

5. The direct application method provides for a reduced amount of total heat release and less damage to the sofa as compared to the same tests conducted using the indirect application method.

RECOMMENDATIONS:

It is recommended that additional research be conducted to develop performance criteria for evaluating the ability of Class A foams to suppress and/or prevent ignition of ordinary combustibles. It may also be desirable to develop a method of evaluating a Class A foam in combination with proportioning and foam generating equipment.

Additional tests should be undertaken to further quantify the firefighting performance and overall improvement in operating efficiency when Class A foam solutions are used with hand hoselines for structural firefighting.

Additional research should also be conducted to determine the optimum tactical approach in manual fire combat for different Class A foam expansion ratios, drain times, and bubble structures.

APPENDIX D: GLOSSARY

The following definitions may be helpful. All are included as they would be applied to the use of class A or compressed air foam systems.

Biodegrade--Decompose to inert or basic ingredients, usually by microbial action.

Burnback resistance--The ability of the foam to resist direct flame impingement.

Class A fire--Fire in ordinary combustible solids such as wood, fabric, paper, or organic materials.

Class B fire--Fire in flammable liquids, gasses, or chemical fires such as burning plastic.

Class A foam--Foam intended for use on class A fires.

Class B foam--Foam intended for use on class B fires.

Compressed air foam systems (CAFS)--A high energy foam delivery system in which compressed air is mixed with a solution of water and class A foam concentrate to produce an expanded, aerated class A foam.

Concentrate--Chemical product that is mixed with water to produce a foam solution.

Concentration--The amount of foam concentrate in a foam solution, usually expressed as a percentage.

Drain time--The time it takes for the foam to break down as the water separates (drains) from the aerated foam bubbles when they break.

Eductor--A mixing system that uses the venturi vacuum effect of flowing water to draw foam concentrate into the water stream, producing a foam solution.

Expansion ratio--The volume of foam formed from a given volume of foam solution; a 10:1 expansion ratio indicates that 10 gallons of foam are formed for every one gallon of foam solution.

Foam--The expanded mixture created when a foam solution of water and concentrate is mixed with air.

High energy system--A foam delivery system where pneumatic energy from an air source such as an air compressor is added to the hydraulic energy from the water pump to both aerate the foam solution and to propel the foam.

High expansion foam--foam designed for air-to-foam ratios of 200 parts air to each part foam solution or higher.

Inductor--A mechanism which allows a regulated amount of foam concentrate into a hose or pump, forming a foam solution.

Low energy system--A foam delivery system where hydraulic energy from the water pump propels the foam solution and the solution is aerated at the nozzle to form foam.

Mix ratio--The ratio of liquid foam concentrate to water, usually expressed as a percentage.

47

Nozzle-aspirated foam system--A low energy foam producing system where air is mixed at the nozzle with the foam solution of water and concentrate to produce foam.

Proportioner--Electronic or mechanical device which pumps foam concentrate into a hoseline or pump at a specific rate or concentration. See inductor.

Solution--Foam concentrate mixed with water. When air is added this solution becomes foam.

Slug flow--In CAFS, when a foam solution does not contain a rich enough amount of foam concentrate to promote the mixture of air and water, pockets of water and air (slugs) in the hoseline result, which can adversely affect water streams and cause violent nozzle reaction.

Surface tension--The attractive force in the surface of a liquid, especially water, caused by the affinity between the molecules in the liquid. This attraction leads to a decreased surface-to-mass area and inhibits flow of the liquid.

Surfactant--A substance which acts to reduce the surface tension of a liquid; a wetting agent.

Wet water--Water with added chemicals (surfactants/wetting agents) that tend to increase the ability of water to spread and penetrate into fuels by reducing the surface tension in the water.

Wetting agent--An additive to water that reduces the surface tension of the water. See surfactant.

Bibliography

Abernathy, David, "There are more than CAFS in Texas... and that's no bull," *The California Fire Service*, March 1990.

"An Operational and Tactical Guide to Ground-applied Foam Applications," United States Department of the Interior, date unknown.

Bethune, Frederic, "A Study on the Effectiveness of Ground Applied Compressed Air Foam Systems (CAFS) in Wildland Fire Suppression in Alaska," Applied Research Project, Executive Fire Officer Program, National Fire Academy: December 1990.

Blankenship, Paul, "Foam: Should We Use It?" *American Fire Journal*: July 1991.

Boise Interagency Fire Center Foam Project, "Issue Paper", March 1990.

Carringer, Rod, "Class A Foam: Awareness and Operations Level Workbook and Glossary."

Carringer, Rod, "Foam Facts: Everything you wanted to know about "Class A" Foam," *American Fire Journal*: November 1990.

Colletti, Dominic, "Class A Foam: Q & A," *Firehouse*: March 1993.

Davis, Larry, "Class A Foam," *Firehouse*: April 1991.

Duncan, Sam, "Evaluation of NDI Compressed Air Foam System (CAFS) Applied as a Retrofit," US Army Tank Automotive Command Research, Development and Engineering Center, August 1994.

"Engineering Analysis of Threshold Compressed Air Foam Systems (CAFS)," United States Department of Agriculture: United States Forest Service Technology and Development Program, October 1987.

"Foam vs. Fire: Class A Foam for Wildland Fires," National Wildfire Coordinating Group: October 1993.

Jones, Kenneth, "Put the white stuff on the red stuff," *Fire Chief*: March 1995.

Liebson, John, "Is Water Really the Answer? New Foams May Be Better, Part II," International Society of Fire Service Instructors, *The Voice*, July/August 1990.

"Manufacturer Submission Procedures for Qualification Testing of Wildland Fire Chemicals," United States Department of Agriculture: United States Forest Service Technology and Development Program, December 1989.

Marx, Martin, Mark Harper and Todd Halter, "Introduction to Quantitative Modeling of Firefighting Foam," Computer Integration and Literacy Institute, Boise, Idaho: 1988.

McKenzie, Dan, "Pump Flow Testing," unpublished, August 1990.

National Wildfire Coordinating Group, "Foam Applications for Wildland and Urban Fire Management," Volume 1: Number 1, 1988.

National Wildfire Coordinating Group, "Foam Applications for Wildland and Urban Fire Management," Volume 1: Number 2, 1988.

National Wildfire Coordinating Group, "Foam Applications for Wildland and Urban Fire Management," Volume 1: Number 3, 1988.

National Wildfire Coordinating Group, "Foam Applications for Wildland and Urban Fire Management," Volume 2: Number 1, 1989.

National Wildfire Coordinating Group, "Foam Applications for Wildland and Urban Fire Management," Volume 2: Number 2, 1989.

National Wildfire Coordinating Group, "Foam Applications for Wildland and Urban Fire Management," Volume 2: Number 3, 1989.

National Wildfire Coordinating Group, "Foam Applications for Wildland and Urban Fire Management," Volume 3: Number 1, 1990.

National Wildfire Coordinating Group, "Foam Applications for Wildland and Urban Fire Management," Volume 3: Number 2, 1990.

"Report of Class A Foam Tests," Department of the Army Belvoir Research, Development and Engineering Center; Underwriter's Laboratories Project 93NK24320, February 1994.

"Report of the National Class A Foam Research Project: Phase II - Structural Fire Fighting," National Fire Protection Research Foundation; Underwriter's National Laboratories, Inc. Project 94NK14167/NC987, December 1994.

Rochna, Ronald, Clarence Grady and Paul Schlobohm, "A Performance Test of Low Expansion Nozzle Aspirated Systems and Wildland Foam," United States Department of Interior - Bureau of Land Management.

Rochna, Ronald and Paul Schlobohm, "A Report on Ground Applied Foam," March 1988.

Rochna, Ronald, Paul Schlobohm and Alan Olsen, "Proportioners", unpublished research paper, date unknown.

Wallace, Gary, "Firefighting Foam: Compressed Air Foam Systems are Changing the Potential for Foam Uses in Class A and Structural Use," Firefighter's News, December-January 1989.

Weider, Mike, "Firefighting Foams: Understanding the Basics," NFPA Journal, May/June 1995.

www.ingramcontent.com/pod-product-compliance
Lightning Source LLC
Chambersburg PA
CBHW081225170526
45165CB00009B/2952

* 9 7 8 1 4 8 2 6 7 5 9 9 3 *